SANCTUARY LOST

How Our National Wildlife Refuges Became
Hunting Grounds

And why hunters have it wrong

❧

JEREMY SCHMIDT

Crowsnest Books

2402 Kendall Ave, Ste 100
Madison, WI   53726
USA

ISBN: 978-1-881480-17-4

CONTENTS

AUTHOR'S NOTE

I am not opposed to hunting, and do not intend
this book to appear as an argument against it. I have
hunted. I have killed for meat. I have cherished my
time in the forests and meadows, sometimes carrying a
gun, always trying to pay attention and learning what I
could about the natural world.

My focus here is on national wildlife refuges. I
want to raise questions about the propriety of hunting
on refuges in general, and particularly the ongoing
expansion of hunting on refuges—an expansion that
seems in some cases a willy-nilly, rushed attempt to
pry open the last acres of protected land, as if the
opportunity might not exist in future.

I also want to correct the mistaken idea that
hunters are the country's major supporters of wildlife.

7

That is a damaging, exclusionary myth. In fact, it's a lie.

I want to set the financial record straight. Hunters claim that they provide most funding for wildlife conservation. They don't.

As for my personal boundaries, I do oppose some kinds of hunting—hunting of certain species, in certain places, and of certain individual animals. For example, I believe swans should never be taken as game, even though they are classified as such in many states. I feel that most wildlife sanctuaries, whether federal, state, or local, should be off limits. They should be refuges in reality, not only in name. When it comes to individual animals, some are in a class of their own. Individuals that become well-known, easily recognized, and locally beloved; others that play an essential role in the social structure of their community (a lead cow in a group of elk, for example)—these should be protected no matter what their species. Admittedly, these distinctions are not easy to define. But I feel they should be contemplated.

Further, I do not accept some justifications for killing things. That a population of animals can survive the death of some does not by itself make it acceptable to kill them. "We manage populations, not individuals" is a basic principle of wildlife management. It's a pragmatic approach, and a convenient one. But it crumbles as a justification. The view of animals as faceless components of a living commodity requires a sort of data-centric blindness in which population data mean more than the creatures being measured. You can't see individuals if you believe they count for nothing more than, say, a specific gallon of water flowing in a river. If you believe that, then an individual deer has no more worth, relative to the herd, than the gallon of water does to the river; the river, or the herd, is the important thing.

I agree that water is a commodity. In the case of animals, however, I know better. Any observant person does; many choose to ignore it. After all, believing that individuals matter puts one at the risk of feeling empathy, which in turn changes one's perception of killing. In one view, the commodity perspective, killing has no impact beyond a statistical measurement. In the other, killing takes a life, a single life, the life of one breathing, sentient creature.

Any hunter with a spaniel, or a labrador, or even a house-dwelling Lhasa apso with ribbons in its hair makes that distinction all the time. Dogs have a different status than deer. They are individuals. Friends. Empathetic companions. A hunter never refers to dogs in the commodity sense, as dog, the way he speaks of bears as bear. That he could never hunt dogs goes without saying.

To kill other animals in a mindful manner—the only respectable manner in my view—demands acknowledgment of the individual life taken. It's a matter of honesty. To do otherwise is a kind of denial. Killing an animal has always had a profoundly humbling affect on me. It comes with regret, gratitude, and even a tinge of shame that requires careful re-balancing. For the life of me, I cannot understand the bravado displayed by so many hunters after inflicting death on a fellow creature.

If I substitute the words "wild animals" for "land" in the quote from Aldo Leopold that opens this book ("When we see land as a community to which we belong, we may begin to use it with love and respect,") I find that the concept shifts, and I can no longer include Leopold's verb "to use." *Using* wild animals strikes me wrong in a way that using land does not; it feels like a violation of the community relationship that Leopold felt was central to a responsible life. The statement needs reconstruction. It becomes, "When we see wild animals as members of a community to which we belong, we may begin to reconsider what it means to treat them with love and respect." The replacement words encourage a shift from commodity-thinking into something different and more personal. I don't *use* my dog; then how can I use a wild creature?

Finally, regarding the benefits of hunting as a

way of connecting with nature, I don't dispute the potential. A mindful hunter—one who ponders and admits the significance of taking life, who approaches the act of killing with humble reverence—confronts a profound and ancient experience. I have respect for such hunters. I know several. They are, sad to say, not common.

We should all be mindful of nature. I believe it is crucial for our health and the health of our land that people know about wild country, and that they learn to appreciate wild creatures in both romantic and unromantic ways; that is, both emotionally and objectively. Hunting is one way of engaging with wild country, but hunting is only one way of many. The same lessons, and many profoundly different ones, can be learned without killing. Some can be learned only without killing. Most Americans who appreciate nature by non-lethal means already know this. In a world of diminishing wildness, it is an important lesson—a lesson that begs consideration on our national wildlife refuges turned hunting grounds.

SANCTUARY LOST

It begins with a declaration of high ideals. On the welcome page of its website, the National Wildlife Refuge System prominently displays the following:

> "When we see land as a community to which we belong, we may begin to use it with love and respect."

> *- A Sand County Almanac*

The writer was Aldo Leopold, pioneer conservationist, natural philosopher, and revered father of modern wildlife management. His words reflect the notion that by honoring the land and its creatures, we honor ourselves and the roots of our existence.

The website continues with a list of governing principles. The first: "We are land stewards guided by

Aldo Leopold's teachings that land is a community of life and that love and respect for the land is an extension of ethics."

Leopold was writing about land in general, from cities all the way to deep wilderness. But nowhere is the sentiment better suited than to the vast assemblage of America's wildlife refuges. From Alaska to Florida, and into remote regions of the Pacific Ocean, more than 500 refuges add up to the world's largest and most comprehensive system of public lands dedicated to the conservation of fish, wildlife, and plants. It's a uniquely American accomplishment. No other country has done anything quite so ambitious and forward-thinking in recognizing the importance of wild habitat to the creatures who rely on it, and to the citizens who look to it as their national heritage. This is Leopold's "community to which we belong."

As you would expect, the system offers a range of loving and respectful uses — watching birds, learning wildflowers, canoeing through wild estuaries, photographing a misty wetland, walking an ocean beach, listening to owls in the night. And to coyotes howling, gators bellowing, cranes clattering in the sky, and bull elk bugling in the frosty autumn dawn.

Americans visit by the millions. The numbers increase every year. We hunger for these places teeming with wildlife. We own them, all of us. They are ours.

Our property, our heritage, our history. Yet by law, people must step aside for the natural residents. On refuges, all human activities take second place to the critters. The U.S. Fish and Wildlife Service, which manages the system, proudly declares it thus: "Wildlife comes first."

Which only seems right, in places called refuges. Bird watchers, as much as they love seeing the plovers and curlews and cranes, are not allowed to disturb waterfowl on their nests. Vehicles are kept off the beach in North Carolina to avoid crushing the eggs of shore birds. In Oregon, volunteers help take down barbed wire fences to make life easier for fast-running pronghorn antelope. On refuges, you can help with bird banding, guard a sea turtle nest, fight invasive plants, and in general do all sorts of things to learn about, appreciate, and help protect wild animals.

And, you can kill them.

You can shoot them. You can catch them in leg traps. You can pierce them with arrows. You can take their dead carcasses home and hang their heads on your wall.

Across the nation every fall, and in some cases at other times of the year, hunters descend on the refuges. They take wood ducks, tundra swans, snow geese, elk, antelope, deer, moose, bears, coyotes, bighorn sheep, mountain lions, sandhill cranes, and so much more. In essence, any creature that can be shot on public land in America — not just game animals, but also predators and what some states still officially call varmints — can be shot on a refuge, somewhere.

Refuge: The word means a place of safety from danger. How can it be that wildlife is not safe on a wildlife refuge?

☙

Refuges didn't start out this way. The first ones, created more than a century ago, were true sanctuaries, off-limits to just about any human use.
Their creation was inspired by terrible events — the brutal nationwide slaughter of almost any wild animal that walked, flew, or crawled within range of a weapon. It was a grand wastage that nearly destroyed some of our most cherished species. Bison, trumpeter swan, pelican, flamingo, ibis, grizzly bear, Stellar's sea lion, sea otter, and manatee are a few that came within existential inches of vanishing forever. Others did disappear. We will never again see passenger pigeons, great awks, Carolina parakeets, Eastern elk, California golden bears, and others.

The first refuges were meant to give wildlife places of safety. They were islands of sanctuary at a time when hunting regulations were few, and hunters showed no sense of restraint. On the contrary, hunters found the very idea of refuges offensive. To exclude hunting from even small patches of ground? It was un-American. A trampling of state's rights. A denial of personal freedom.

Hunters resisted refuges but they couldn't stop them. In 1903, President Theodore Roosevelt laid the cornerstone for today's continent-spanning network of refuges. The cornerstone — scarcely a pebble, it was so small — was Pelican Island, an obscure three-acre patch of mangroves off the coast of Florida. Roosevelt had a particular affection for pelicans, egrets, flamingos, and others of our spectacular waterbirds. He was appalled at their slaughter by plume hunters serving the demand for women's feathered hats. When he asked his advisors to find a legal authority under which he could designate protected land for birds, they told him no such law existed. There was a law covering establishment of forest reserves; but nothing for birds. Well then, he famously asked, was there a law that prevented him from doing so? There was not. So, with a simple declaration using the vested power of his office, he established the nation's first federal bird reservation. He followed it with dozens more, the first in what became the vast system of refuges we enjoy today.

Roosevelt did not act alone; but neither was he responding to any sort of significant political pressure.

His collaborators and supporters were a small band of forward-thinking conservationists and scientists. Some of them, like Roosevelt, were hunters. Some were not. They worked quietly and stepped lightly, knowing that the national sentiment was against them.

What would they think, these foresightful few, to know that their sanctuaries would eventually become the nation's favored hunting grounds? That what they saw as inviolate refuges for wildlife, where hunters could not so much as set foot, would become the opposite? Today, a refuge closed to hunters is an exception. Most units in the system not only permit shooting but actively promote it, in many cases giving hunters precedence over other users. During hunting season, sections of some refuges — sometimes entire refuges — are off-limits to anyone *not* carrying a weapon. At those times, bird-watchers are forced to yield to bird-shooters. Photographers hoping to capture a bull moose in his autumn glory are trumped by a hunter who wants to remove it from the scene forever. A thousand viewers could appreciate that one animal, repeatedly, every autumn over the course of his natural years. Yet a single person is allowed, even encouraged, to kill him. And then officially praised for doing so.

For their part, hunters are militant about the issue. They see hunting as an essential heritage that springs from a love of the outdoors and a special connection with their wild quarry; indeed, some see it a loving and respectful use. They trace their roots to a pioneer past when hunting provided an essential touchstone to nature and the understanding of the human role in natural systems. They also claim that hunting on refuges is a legacy bought and paid for by their own conservation efforts, thereby earning them a sort of founders' right. Without hunters, they say, there would be far less wildlife for the refuges to protect.

That's not true. But it's not entirely false.

A BRIEF HISTORY OF
TARNISHED IDEALS

To understand how this happened — that is, how refuges became hunting reserves — requires digging into a complicated history.

The refuge system as it stands today is one of America's great achievements. As a public endeavor, it ranks on a par with the national parks system. In fact, it's nearly twice as large, but remains far less visible to the general public.

Although called a system, it's actually a sprawling disarray of holdings created under a variety of legal actions with different and sometimes conflicting purposes. Currently it covers more than 150 million acres in 553 primary units (the parts officially called national wildlife refuges) plus 38 wetland management districts and thousands of small holdings. Stretching from the tip of the Aleutian Islands in Alaska to the Tortuga Islands in Florida, it takes in virtually the entire span of the continent's wild habitats. It includes far-flung islands in the Pacific Ocean: Midway Island, Kingman Reef, Wake Atoll, and others. Also part of the system are undersea volcanoes not visible from the surface, and even the black, unexplored depths of the Mariana Trench.

The U.S. Fish and Wildlife Service manages it all, and proudly counts among its residents more than 700 species of birds, 220 species of mammals, 250 species of reptiles and amphibians, and more than 1,000 of fish. Many species are migratory. Turning up in backyards, farm fields, county parks, school forests, national parks, and other lands, they enrich our lives and delight our senses.

The idea of protecting places for wildlife goes back centuries, with examples found in many countries. Yet nowhere has the idea become so well established, so effective, and so popular, as in the United States. Although earlier presidents established protected areas (Ulysses Grant set aside the Pribilof Islands in 1868, for northern fur seals, and Benjamin Harrison reserved Afognak Island for timber and fish-culture in 1892), founding credit for the refuge system usually goes to the man called our "conservation president," Theodore Roosevelt.

When he declared that Pelican Island was "hereby, reserved and set apart for the use of the Department of Agriculture as a preserve and breeding ground for native birds," Roosevelt was taking advantage of an obscure statute allowing a president to create forest reserves. Never mind that the island had no real forest. It was a favored nesting ground for brown pelicans, herons, roseate spoonbills, egrets, and other coastal birds. Roosevelt had a particular affection for birds, and wanted to save a few from market hunters who at the time were slaughtering millions for feathers to decorate women's hats; also from joy riders in boats who shot seabirds out of the sky simply for the pleasure of watching them fall.

It was a time of appalling wastage. America's wildlife was under attack by sport hunters, market hunters, subsistence hunters, and many who shouldn't be called hunters at all, because they had no interest in the spoils.

In those years, hunting was a matter for the states to govern, but few regulations existed and enforcement was weak or non-existent. With essentially no limits on what could be killed, and by what means, and when, and how many at a time, the destruction of America's wildlife had reached a critical stage.

Forty years after Pelican Island, Ira Noel Gabrielson, the refuge system's first director and an eminent conservationist, wrote a history of the refuges up to the time of World War II. He described how close the country had come to losing its wildlife, and how precarious the situation remained. "The tale of greed and exploitation," he wrote, "has been told many times." And a gruesome tale it was. He adds, "The story of its halting in time to save remnants… is still too recent to be completely unfolded. The achievement of the once-small but steadily growing band of valiant spirits, further visioned than their contemporaries, is an inspiring tale."

In telling that story, Gabrielson wrote about policy without personality. He scarcely mentioned the individuals who comprised that band of valiant spirits—not only Roosevelt, but William Hornaday, John Muir, George Bird Grinnell, Frank Chapman, and others. Roosevelt, it must be acknowledged, occupied the center of political power. But he did so with the support, inspiration, and aid of a small, fervently dedicated group of believers.

The story of those people, the more personal story, is filled out admirably by historian Douglas Brinkley in his book *Wilderness Warrior,* a biography of Roosevelt as pioneering conservationist. Some modern critics, opposed to hunting in general, have faulted Roosevelt for being a hunter. They accuse him of wanting to protect animals only so he and his ilk could blow them to bits. They claim that his motives were impure and that he should not be given equal standing with vegetarian purists like John Muir. Brinkley, however, makes it clear that T.R. was a far more sophisticated and broad-minded lover of nature than those critics would admit. He describes the president as a man who "held a romantic view of the planet, a belief that *Homo sapiens* had a sacred obligation to protect its natural wonders and diverse species." It wasn't just game species that he valued.

Since boyhood, Roosevelt had felt a passion for the natural world. He had traveled. He had seen the best of wild America. From first-hand experience, he knew what the country stood to lose. And as President, he could do something about it.

Pelican Island was a small first step toward a grand achievement. Roosevelt wasted no time, following up with 51 more federal bird reservations and four big game reserves before leaving office in 1909. He was also responsible for setting aside huge tracts of other wild land, with results that today are astonishing in their breadth and scale. Taken all together, the national parks, forests, monuments, refuges, and other preserves created under his signature—if we count acreage added by successor administrations to sites Roosevelt established—cover 234 million acres. This amounts to about 10% of all American land including Alaska.

What Roosevelt and his allies accomplished built the enduring foundation for wildlife conservation in America.

Yes, he was, famously and proudly, a hunter. He killed many animals, from the American West to the savannas of Africa. He believed that pursuit of game in the rugged outdoors was an essential part of being American, and that his time stalking the wild — especially on his beloved Dakota plains — had fundamentally influenced his personal character. He was a hunter until the end of his life, as were many of his close friends and advisors (John Muir being a noteworthy exception and critic). But in those crucial decades of the early conservation movement, Roosevelt was an unusual hunter. He held a view of nature that, in its time, was broad, sophisticated, and shared by a select few others.

Brinkley writes that Roosevelt "had a keen sense of the importance of what would come to be known as biological diversity and deep ecology.... To him the destruction of pelicans — and other non-game birds — was emblematic of industrialization run amok." Zoologist William T. Hornaday, a contemporary and friend of Roosevelt, wrote in 1913, "He aided every wild-life cause that lay within the bounds of possibility, and he gave the vanishing birds and mammals the benefit of every doubt."

As Roosevelt envisioned it, the refuges he created would serve as fully protected zones. They would be storehouses and breeding grounds. They would sustain wildlife in the face of whatever exploitation or destruction might occur in other places. They would protect not only species of interest to hunters, but sandpipers, plovers, gulls, terns and many other birds that no one thought of eating or plucking. They would also provide sanctuary for the small mammals, reptiles, insects, and plants associated with each location. Roosevelt understood the importance of habitat and diversity. Rather than draw lines around individual species—banning the killing of egrets, for example—he did better by protecting the natural areas that sustained them.

Or rather, he began a long and often compromised process.

The simple approach was never practical, and Roosevelt probably knew it from the outset. That is, if animals were allowed to function in peace according to nature's own rules, without interference from people, they should thrive. It should be enough to leave them alone in strictly protected enclaves that would serve as nurturing grounds—storehouses of nature to indemnify the entire country against the irrecoverable loss of its precious wildlife heritage.

There were two reasons why this approach wasn't

enough. First, there needed to be a system, not just a scattering of isolated pockets established wherever an opportunity made it possible. You couldn't protect wildlife piecemeal. Migration routes, surrounding land uses, seasonal variations, and a host of other factors had to be considered. Migratory birds needed protection not only in their nesting grounds, but also in their wintering territory, and along the flyways that stretched between them. To make this happen, refuges should comprise a mosaic whose units reinforced and complemented each other. Doing it right would require continent-wide planning backed up by comprehensive scientific knowledge that didn't exist in the beginning, coupled with strong political support.

Second, the system had to be managed and paid for. Much of the needed land already belonged to the public, but a lot of the most valuable had to be bought or otherwise acquired from private owners. All of it needed protection—primarily from hunters in the early years, but also from other threats large and small. Some critical areas, particularly wetlands in the 1930s, were so badly degraded by careless development that they needed significant restoration. Species needed restoration, too: dozens were saved from extinction only by intensive efforts involving captive breeding, restocking, feeding, habitat manipulation, and coordination with surrounding land owners. Wildlife needed more than passive protection.

At the turn of the 20th Century, wildlife was not yet a federal responsibility. Its management, if any, was generally seen as a matter for state or even county government. Most states had no hunting regulations whatsoever. The federal government had no practical experience in wildlife management; nor did it have a clear mechanism or a budget for enforcement of whatever rules it might create. As Gabrielson put it, refuges existed "in name alone, for [refuge wardens], with insufficient funds and with inadequate public support, could do little to protect them from poachers." The first wardens were private citizens hired with private money. In the case of Pelican Island, the warden was a local resident with a passion for protecting birds. His small salary and equipment was paid for not by the government but by the Audubon Society.

It was decades before the U.S. Fish and Wildlife Service (created from other agencies in 1940) took charge of protecting and managing the refuges. They were not even called national wildlife refuges until 1942.

Through it all, one thing remained clear. Starting with Roosevelt and continuing until after World War II, the intent of federal legislation and executive orders was that wildlife refuges would provide comprehensive sanctuary. Animals on a refuge were protected from human disturbance of any kind — roads, buildings, machines, mining, drilling, farming, recreation, and any other intrusion that disrupted the natural process of their lives. That certainly included protection from being shot to death by hunters.

The concept of complete sanctuary, established at the outset, was recognized and accepted by progressive hunters like John B. Burnham, president of the American Game Protective Association, an organization of hunters. In 1919, Burnham wrote an editorial in the association's bulletin. He proposed the creation of a nationwide system of "public shooting grounds" to benefit future generations of hunters. In the same article, he echoed and reinforced the established view of inviolate reserves: "With the public shooting grounds must come more reserves where the birds should have absolute protection, for as the country becomes more settled, shooting would become impossible without them." In essence, he proposed a grand system of inviolate refuges partnered with hunting areas in a way that retained the sanctuary status of the refuges while also formalizing hunter access to other areas. The system, both hunting and non-hunting parts of it, would be paid for through sales of federal waterfowl hunting licenses.

His idea hit a nerve, and gathered strong support

across the country. It also aroused opposition from states' rights advocates who believed that the federal government had no authority over migratory wildlife. Opposition also came from various groups, including hunting clubs, who denied that a wildlife problem existed in the first place; and from people on the other side of the issue who took offense at the coldly mercenary idea of pairing refuges with places designated for shooting the very animals being protected. Ironically for the latter folks, killing grounds and nurturing grounds were eventually not separated by even a fence line. Instead, they became one and the same.

For a decade, the argument spiraled through Congress in a series of failed bills, and ended with a compromise, the Migratory Bird Conservation Act of 1929, which provided for the purchase or lease of sanctuaries for migrating birds. These would be paid for by general appropriations (not hunter fees), and there were to be no adjacent shooting grounds. The law defined migratory birds according to the terms of international treaties, the most important being the Migratory Bird Treaty Act of 1918 which carried the first significant federal restrictions on hunting. The 1918 law recognized the shared interest by various entities in species that crossed state and international boundaries. Geese might nest in the Canadian arctic, but they wintered in the south, and everyone along the way has claims on their management. The sanctuaries authorized by the 1929 law would help meet treaty obligations. They would protect not just game birds but almost all birds that occur naturally on United States territory, from robins to ibis to wood ducks. Over the years, this treaty and other international agreements have had an important influence on refuge establishment and location.

Although the Migratory Bird Conservation Act did not appropriate funds to accomplish its ends, it did reinforce the concept of inviolate sanctuaries and the federal role in establishing them.

As drought gripped the nation in the 1930s, the quality of habitat became a serious concern. Wetlands were shrinking, demands on the water supply for irrigation grew, and the need for conservation efforts became ever larger. If waterfowl were to recover from the over-hunting of earlier decades, a costly program of wetland acquisition and rehabilitation was urgently needed, along with a way to pay for it. To help on the money side, Migratory Bird Hunting Stamps—better known as duck stamps—were introduced in 1934. Every waterfowl hunter in the country was required to buy a duck stamp (the first ones cost a dollar) and affix it to his state-issued hunting license. Revenues were allocated for acquisition, in the words of the law, of "inviolate migratory bird sanctuaries." The stamps did not buy hunters physical access to hunt on refuges. They contributed to them, and got to shoot the waterfowl that migrated out of them, but the refuges themselves remained zones of total safety.

In 1940, President Franklin Delano Roosevelt further strengthened the status of many refuges. He signed an executive order that changed the names of 193 sites formerly called "reservations" to "refuges" on which it was unlawful "to hunt, trap, capture, willfully disturb, or kill any bird or wild animal... or to enter thereon for any purpose, except as permitted by law or by rules and regulations of the Secretary of the Interior." In the same year, the U.S. Fish and Wildlife Service was established from existing agencies.

This was strict protection with a clear purpose. Refuges were not parks. They were not recreation areas. They were for wildlife. They were certainly not hunting grounds.

But even then, the standards for protection were beginning to change. America was a different place after World War II. With more leisure time and money for recreation, people turned to the outdoors — to parks and forests and, inevitably, to the beautiful landscapes of the refuge system. Pressure grew for public access. At the same time, the refuge system was hurting for operating funds. To policy makers, the tension between demand and funding presented an opportunity. A single answer could help resolve both needs, but only through abandonment of long-standing principle. The result, in the eyes of many, was nothing short of betrayal.

It came in 1949. An amendment to the Duck Stamp law doubled the cost of stamps to $2.00 and, partly in justification for the increase, made a radical break from precedent. Congress opened the gates to hunting. The amended law allowed up to 25% of acreage in previously "inviolate sanctuaries" to be designated as "wildlife management areas." Management in this case meant hunting.

There's a curious but important twist in how this happened. As first introduced, the bill proposed hunting only in newly acquired areas. The idea was to use the increased hunter dollars to buy new places for hunters to hunt, while maintaining and supporting the traditional status of existing, inviolate refuges. In this, there were shades of Burnham's old idea of pairing refuges with shooting areas. However, by the time it was signed into law, congressional committees had modified the concept, applying the 25% rule to all refuges. Whether new or old, whether bought with hunter dollars or not, the policy affected them all. The camel's nose was in the tent. The founding principle of the refuge system was cracked. Hunters had broken through the gates.

Nine years later in 1958 Congress amended the law again, increasing the permitted hunting portion from 25% to 40%, while also allowing duck stamp money to buy new tracts called Waterfowl Production Areas. This was a new category of refuge holdings. Unlike other units in the system, WPAs could be 100% open to hunting.

It was a sort of political ransom. Refuges traded wildlife in return for legislative and financial support. Meanwhile, hunters who paid for duck stamps were encouraged to think that they were buying the rights to collect a toll—in the form of wild lives—on refuge lands.

It has become an increasingly large toll. Under current law, hunting has been further elevated and vigorously promoted to the point that some 97% of refuge land—all types of refuge land—can be opened to hunting.

They can, but not all refuges do. Some allow hunting only under highly restrictive limits. Others remain inviolate to hunting, and a few are off limits to any public use whatsoever. The most restrictive include highly sensitive nesting grounds like Navassa Island off Puerto Rico, and the western end of the Hawaiian Island chain (of which Midway Island is a part).

These last ones, these most strictly protected sanctuaries, illustrate an important policy difference between wildlife refuges and other public lands like national parks. It's a simple and extremely powerful distinction. Refuge lands outside of Alaska are *closed to human use* unless specifically opened. On top of that, they cannot be opened unless the use in question is deemed compatible with their purpose and mission. National parks, in contrast, operate under the reverse policy. Parks are open "for the benefit and enjoyment of the people" unless closed for reasons of resource protection. In both cases, protection of the resource comes first but the path of management is different. A national park wanting to deny or limit use must show that people are causing damage. A wildlife refuge has to show the opposite—that a given use will *not* cause damage—or the use should not be allowed.

Such is the theory. In fact, the story of refuge management is a pile of contradictions. Refuges were created under a hodgepodge of acquisition methods that often involved concessions to competing interests like mineral rights, water rights, grazing, and other uses. In some cases, the establishment documents for a refuge specify purposes that take precedence over, and may even be in conflict with conservation. Some refuges are subject to shared management by the U.S. Fish and Wildlife Service and the military or other agencies. Yet the overarching principle has remained the same, ever since Pelican Island: Wildlife comes first. Human needs and demands are secondary.

In 1962, the Refuge Recreation Act attempted to clear up the welter of contradictions. Public recreation could be allowed on refuges, but only if specific activities were "not inconsistent" with the primary objectives for which each area was established; and only if adequate funding existed for management of those activities. The requirement of being "not inconsistent" is also referred to as "compatibility."

It's a squishy, ill-defined standard, difficult to codify, much less maintain. Over the years, the refuge system has had to cope with repeated demands for exploitation from voices willing to see compatibility in almost any use. One of the worst interpretations came in the early 1980s, when Secretary of the Interior James Watt pushed for any and all economic uses, from logging to mining to oil and gas leasing. These should be allowed, he proposed, if there would be "no significant adverse impact in the present and the lack of irreversible effect in the future." In other words, if damage could be fixed later on, or merely said to be fixable at some time in the future, then any adverse impact was acceptable.

Mr. Watt did not prevail, and the battles over compatibility continued, as lawsuits flew from all sides and various attempts to pass reform legislation arose and failed. The issue only grew hotter, until it eventually spurred passage of the landmark National Wildlife Refuge Improvement Act of 1997. This legislation was intended, among other things, to resolve the issues of compatible use and provide a guiding legislative foundation for the system as a whole.

Signed by President Clinton in 1997, the law retained and enshrined conservation as the *primary objective* of the refuge system, and declared six wildlife-dependent activities to be compatible but *subordinate* uses of refuge land. The language is important. Those six activities — photography, birdwatching, nature observation, nature education, hunting, and fishing — were given priority over other human uses, but none of the six has priority over the others. There is no requirement that refuges allow any of them, and they must all give way to ecological conservation.

The new law sanctioned the notion that refuges should meet human desires — indeed mandated them. Critics thought of it as a further step toward reducing the refuges to multiple-use playgrounds. Some saw it as a surrender of standards that could only lead to further erosion of wildlife protection. On the positive side, the law reaffirmed in clear language the old founding principle of the refuge system: People come second to wildlife.

There are times on refuges when this is not

obvious; the dead moose being winched out of the willows, her orphaned calf hiding nearby; the pronghorn herd, panicked by having been shot at, fleeing to the winds at the approach of a person with a camera; the official signs telling non-hunters to stay out because a given area is reserved for people with guns. And it's the people with guns who have the momentum, as demonstrated by a September 2013 rule change proposed by the U.S. Fish and Wildlife Service, "to add 6 national wildlife refuges (NWRs) to the list of areas open for hunting and/or sport fishing, add new hunts at 6 refuges, increase the hunting activities available at 20 other refuges, and increase fishing opportunities at two refuges, along with pertinent refuge-specific regulations on other refuges that pertain to migratory game bird hunting, upland game hunting, big game hunting, and sport fishing for the 2013-2014 season."

The founding principle is subject to all manner of corrosion, political pressure, varying interpretation, and outright abuse, but it survives nonetheless. People come second to wildlife.

THE PLACE OF HUNTING

Federal law has given hunting a sanctioned place on wildlife refuges. It's an established fact, but one still open to question. How big a place should hunting have?

In the views of many, the ideal system would be a nationwide network of inviolate refuges, coordinated and well-funded. But such a system was never supportable. Politics and fiscal concerns intruded from the beginning, and became ever more intrusive as the system grew. Good habitat and wild land became ever more rare, more valuable, more threatened and—as with all precious things—more desired by exploiters and appreciators alike. Wildlife refuges are in constant need of defense, just as they were in 1903.

As a further irony, hunting on refuges runs against

the demographic tide. Every year fewer hunters take to the field, while non-lethal users turn out in greater numbers. On refuges, unarmed visitors outnumber hunters by about four to one. Yet in the face of this, the Fish and Wildlife Service continues to expand hunting. Some refuges have recently opened new areas to shooters, and added new species to their huntable list of animals. Others are working toward that goal, opposing the efforts of those who would slow them down. An ongoing lawsuit originally filed by The Fund for Animals, and later taken over by the Humane Society of the United States, challenges the opening of new hunting opportunities on 70 refuges. That case remains unsettled.

Meanwhile, hunting has the support of powerful interests. Official refuge publications provide instruction on how to hunt more successfully on refuge lands. Special hunts for youngsters are common. Their stated purpose is to cultivate interest in an activity increasingly less appealing to young people—for the very reason that young people are less interested, as if to say "We need more kids out there shooting ducks, and it's worth spending our limited dollars on getting them to do so."

In 2004, the Fish and Wildlife Service created an office in Washington devoted specifically to the promotion of hunting and fishing on refuges.

In 2010, the refuge system announced the creation of a national advisory council to promote hunting as a valued tradition—a waning tradition that evidently needs the encouragement and financial support of government to continue.

❧

Do hunters deserve more than other citizens? How is it that hunters are permitted to permanently remove animals from a refuge when non-hunters are forbidden to pick wildflowers? Is it right that hunters should have access in the fall—a splendid time to see wildlife—while others must search outside refuge boundaries for a glimpse of animals in their seasonal prime? In cases where all users have autumn access, is it right that hunters should be allowed to put wildlife to flight, scatter herds and flocks, disturb the natural soundscape, and generally alter the experience for everyone else? Should hunting access on refuges be expanded while hunting becomes less popular?

The answers, as presented by hunting groups, are that they were the first conservationists and are still the major supporters of wildlife in America; that they know and appreciate nature better than any other group; that they pay the bills for wildlife management; and that wild animals benefit from being hunted.

Hunters use these arguments to say that they, more than other citizens, have earned a claim to these sacred places and to the wildlife that belongs to all Americans. They've used them with incredible effectiveness, to the point that hunting interests are the dominant voice in wildlife conservation across the country.

The same arguments are used on any public land where hunting is permitted, not just on refuges. But this book is not about those other places, nor is it about hunting in general. This book is about our national wildlife refuges — places designated primarily for the benefit of wildlife. At the very least, the bar of justified killing should be higher on refuges than elsewhere. There is a simple test. If wildlife truly comes first, if the refuge system is to honor its central and often-stated principle, then every hunting action should meet a basic standard. Not that the wildlife can sustain the loss. Not that hunting is an American heritage. Not that one group of people has earned the privilege. In fact, it's not a human question at all. Anyone who wants to allow wild, native animals to be killed on a refuge should have to answer this: How does shooting these animals, in this place, benefit wildlife?

The claims of hunters warrant looking at one at a time.

"WE BUILT THEM."

In March, 2010, Secretary of the Interior Ken Salazar announced the formation of an advisory group to promote hunting on refuges and other public land. Called the Wildlife and Hunting Heritage Conservation Council, its members include the National Rifle Association, Ducks Unlimited, members of the shooting sports industry, and various government agencies involved with wildlife. The idea was for hunting groups to advise government on matters of policy that would enhance and perpetuate hunting as an American tradition.

Salazar's remarks perpetuate a myth. "The early efforts," he said, "of America's hunters and anglers to preserve our nation's wildlife heritage fueled the modern conservation movement and left us the natural bounty we are now entrusted with protecting. In the spirit of Theodore Roosevelt, we are enlisting the help of hunters and anglers to help us confront the conservation challenges of our times."

Politicians and hunting groups have been saying that for half a century and longer, whenever the subject of hunting on public lands comes up. To paraphrase the argument: "We built these places. We supported them from the beginning. Were it not for us, they wouldn't exist. If anyone should reap the rewards, it should be hunters."

Hold your fire, pardner. Let's have a look at that.

Roosevelt, as a devoted hunter and hero of conservation, is the usual figurehead for these arguments. His actions are frequently exploited to point out that it was hunters who sprang into action when the outlook for wildlife was most dire.

It was certainly dire. As early as 1843, John J. Audubon lamented the slaughter of bison as he had witnessed it on the plains. He warned of their extinction—a fate that most observers thought impossible at the time. If they thought about it at all. But by 1900, Audubon's warning had become an almost certain reality. The nation was engaged in a horrific, end-of-the-game shooting spree. No matter how many times we hear them, the numbers are shocking. Buffalo were essentially gone before the turn of the century. So were wild turkeys. Vast flocks of passenger pigeons, billions of birds, arguably the greatest wildlife spectacle in history, would soon disappear forever. Pronghorn antelope, once almost as numerous as bison, were down to a few thousand. So too with bighorn sheep, woodland caribou, and even white-tailed deer. Given the numbers of white-tails today (about 20 million) it's hard to imagine that in 1900, people in the Northeast thought it was remarkable—something worth telling their neighbors—if they saw a deer at all. Deer had been shot into scarcity. Meanwhile, plume hunters were wiping out entire coastal rookeries of wading birds and seabirds. In restaurants, songbirds were commonly on the menu.

Hunters went after anything that walked, flew, or crawled. They used any tool at hand—rifles, nets, fires, explosives, pits, clubs, poison, and even cannons loaded with metal bits like giant shotguns, mounted in small boats to fire on flotillas of waterfowl. They killed for food, for money, and for pleasure. They took plumes, hides, fur, bones, and meat. Too often they took nothing at all, leaving spoils to rot, as if simple destruction were an honorable goal in its own right.

Today, it defies imagination that people could wreak such havoc on wild creatures, but the context of the times was different. Many assumed that wildlife numbers were inexhaustible. Few believed that animals had any rights of their own. Others looked upon wildlife killing as noble work, a necessary step in civilization's advance, and at the very least, an inevitable result of progress. In their thinking, the clearing-away of forests, swamps, and wild animals would lead to a more enlightened way of life; extermination was therefore not the tragic loss of a national heritage, it was the noble future.

In 1900, the very word conservation, as a term referring to the prudent treatment of nature, did not yet exist.

Hunters often distinguish sport hunters of those times from market hunters, putting the blame for wholesale slaughter on the people who shot for money or skins or tongues, who viewed animals as a commodity, and collected wildlife with the same rapacity with which timber barons went after forests, and miners tore up the mountains for gold. Certainly, abuse was widespread. Legions of animals fell to people who had no interest in sport. Yet sport hunters were not above the misbehavior.

William Trefethen, in his book *An American Crusade for Wildlife*, published by the Boone and Crockett Club, quotes an 1884 report in *Forest and Stream* magazine about goose hunting on the Platte River in Nebraska:

"The gunners have so increased in the last three years that the weary goose, coming down from the North, or in the fields to rest or slake its thirst, can hardly find a place out of the range of some one's gun. Blinds line the bars in the stream for one hundred miles so thickly as to preclude all chance of a fair bag."

Trefethen adds that new railroads had brought development into the best waterfowl breeding grounds, and provided a means for market hunters to ship vast numbers of birds to population centers on both coasts. The railroads "also brought in hundreds of sportsmen whose cumulative effect on the waterfowl was as great or greater than the pressure exerted by the comparatively few who shot for profit."

As Gabrielson put it, conservationists were rare. "Even during the period of wildest exploitation there were those who realized the folly… but their voices were little heeded by the hurrying, bustling, white hordes bent on conquering and consuming a continent in record time."

So who were Salazar's hunters? The ones who, as he put it, left us our natural bounty? Surely not the wielders of wildlife cannons. But they also were not the general hunting public.

Rather, it was a select few who raised their voices against the slaughter—a small minority of mostly well-educated city dwellers, hunters and non-hunters alike, who had the money and leisure time to support conservation. The hunters among them admitted to a selfish interest. They wanted their sport. They wanted to save the game before it vanished. Yet the best of them recognized that wildlife and wild places held intrinsic values that went far beyond sport.

Roosevelt was their leader, and exemplar. Coming home to New York after a life-changing experience in the wilds of Dakota, he founded, in 1887, the Boone and Crockett Club, named for Daniel and David, the two iconic pioneer woodsmen. For members, he enlisted hunters who, like himself, believed there was a fundamentally American, character-building value in the pursuit of wild game. There were things to learn about oneself and the surrounding world that only life in the rugged outdoors could teach — important moral lessons that went to the heart of what it meant to be a responsible and upright citizen. But it had to be done according to the rules of what they called "fair chase." Their approach came with a strict protocol that ran counter to the terrible hunting practices common to the time. They would hunt only game animals, and only by methods that gave the quarry some chance of escape. They insisted on scientifically based hunting regulations: seasons, strict bag limits, and limited weaponry. They wanted nothing of egregious slaughter techniques. No killing of deer mired in deep snow or swimming across lakes, no nets, no explosives, no blasting birds from night-time roosts or in rookeries. They disapproved the shooting of song birds and other non-game species. These were all solid progressive ideas, except for one enormous exception. That exception was predators, which they and many others, including people opposed to hunting in general, felt were nothing better than noxious creatures to be removed by any means at hand. It would be a long time before ecologists and wildlife scientists granted predators the ecological respect they deserve; many hunters still can't see it.

A century ago, killing predators was approved and supported by all sorts of people, including those who cared about wildlife and believed that it was in the best interests of conservation to destroy animals that killed other animals. They targeted prominent species like wolves, mountain lions, and bears, but also the lesser actors—hawks, falcons, bobcats, foxes, weasels, and the like. Believers included supposedly well-informed scientists and nature lovers. The Audubon Society, for example, recommended killing hawks and owls. The National Park Service shot and trapped predators, and managed to help exterminate wolves in most of the lower 48 states.

Attitudes changed slowly. In his 1953 book *Round River*, Aldo Leopold wrote a famous essay, a sort of confession, in which he described the killing of a great bear, a lone grizzly who lived on an Arizona mountain called Escudilla. The bear was taken by a government trapper. As a young forest officer and government employee, Leopold had, in his words, "acquiesced to the extinguishment of the bear." Years later, he felt the loss. "Escudilla still hangs on the horizon, but when you see it you no longer think of bear. It's only a mountain now."

Even today, predators occupy a separate territory among wild animals. Despite overwhelming science to the contrary, they are still widely thought to be enemies of other animals, sharing the conceptual neighborhood with what, in some states, are still officially called varmints — creatures so low, so despised, that they can be killed any time of year in any numbers by almost any means. The events that led to the conservation of game animals in Roosevelt's time had little relevance to these reviled species. Until recent decades, with the arrival of laws like the Endangered Species Act and the enlightened support of a better-educated citizenry, the best chance for predators was to benefit indirectly from conservation projects targeting the so-called desirable species (habitat protected for moose and elk also being good for hawks and foxes); or simply to be overlooked by hunters and trappers, because predators are good at being inconspicuous. Current law, improved with regards to some species, nonetheless still permits the taking a variety of animals for nothing more than fur, feathers, or fun, even on many refuges.

By the time Roosevelt became President in 1901, he had built a network of influential conservation-minded supporters in politics and business. With the power of the White House behind them, he and his allies accomplished big things, turning the tide of destruction and planting seeds of conservation that would flourish over succeeding decades. Biographer Douglas Brinkley gives credit to their practical effectiveness: "Many historians now believe that the Boone and Crockett Club—Roosevelt's brainchild—was the first wildlife conservation group to lobby *effectively* on behalf of big game. The club sprinkled the issue of wildlife protection with kerosene, struck a match, and watched it take off. While antihunters sat on the sidelines gabbing about the extermination of the buffalo, Roosevelt and Grinnell popularized the sportsman's code and called for protection of the buffalo in Yellowstone."

This is the banner waved by hunting groups when taking credit for wildlife conservation: Roosevelt and his hunter comrades, battling for the survival of America's natural heritage. It is a powerful banner. America owes gratitude to the people who originally raised it, however select a group they were, and however strong was the opposition raised against them by the general hunting population. They were hunters, and we should thank them.

Yet hunters, historians, and politicians too often diminish the role of the non-hunters and the anti-hunters who, in their view, stood weakly to the side. This is a mistake. It underestimates their influence. Men like Roosevelt were inspired not only by their personal experiences with nature, but also by the likes of John Muir and Frank Chapman — artists, poets, and naturalists, some of them hunters, many of them not. These figures lobbied hard, cheered on their politically connected champions, and kept an eye on the larger (what we today would call the ecological, the biospheric) perspective.

No one did that better than John Muir, who wrote in 1904 a grudging acknowledgment of the awakening sentiment to save some wildlife: "The murder business and sport by saint and sinner alike has been pushed ruthlessly, merrily on, until at last protective measures are being called for, partly, I suppose, because the pleasure of killing is in danger of being lost from there being little or nothing left to kill, and partly, let us hope, from a dim glimmering recognition of the rights of animals and their kinship to ourselves."

It was hard-edged criticism of an activity he deplored, but nonetheless a hopeful tip of the hat to Roosevelt and his band of action heroes.

Muir also expressed a withering irony. Antihunters and non-hunters did not cause the disaster in the first place. It was not the likes of John Muir, armed only with watercolors, notebooks, and an eloquent voice, who took wildlife to the brink of destruction. Nor was it Frank Chapman, the pioneering ornithologist and curator at the American Museum of Natural History. Chapman lobbied tirelessly for the protection of birds and habitat. It was he, foremost among others, who convinced Roosevelt to set aside Pelican Island.

Hunting groups can't take credit for stopping the senseless slaughter at Three Arch Rocks, off the coast of Oregon, where boatloads of Sunday shooters blasted at tufted puffins, guillemots, auklets, murres, and gulls, not one of them to eat or otherwise put to some useful purpose, but simply for the pleasure of seeing them fall. Instead, it was William Finley, a nature photographer scorned by the trigger-happy crowds, and his friend Herman Bohlman, who led the charge. Their photographs entranced Roosevelt, who made Three Arch Rocks a bird sanctuary in 1907.

Trumpeter swans did not reach the edge of

oblivion in the 1930s because their wetlands had disappeared. Loss of habitat—the ongoing disaster that today poses the major threat to wildlife—was not the controlling issue at the time of their great population decline. Trumpeter swans were slaughtered by hunters first. The dust bowl made things critically worse. By 1934, trumpeters had disappeared from everywhere in America except the remote Centennial Valley in Montana, a broad subalpine wetland where fewer than 100 birds managed to hang on. To hear hunters talk these days, you would expect that the local hunt clubs, which maintained shooting camps in the valley, had been crying out for action to save the last of these noble birds. Instead, they opposed creation of the Red Rock Lakes National Wildlife Refuge because it might limit their shooting. They lost. The refuge happened against their objections, and trumpeter swans were saved.

It took a powerful load of catastrophe to finally get the attention of hunters. James Trefethen writes that after decades of unremitting spring-and-fall shooting in the upper midwest, and despite warnings like the one published in *Forest and Stream* in 1884, gunners were running out of targets. In the end, it was climate, not negotiation or education or plain common sense, that forced the issue. In 1899, Trefethen writes, "waterfowl flights plummeted to an all-time low as a searing drought spread across the breeding grounds. It suddenly became apparent that if duck and goose shooting were to be preserved as traditional American sports, drastic steps were needed."

This recognition marked the beginning of better times. A critic might point out — as Muir did — that it was not the intrinsic value of migratory birds, the joy of seeing them in the autumn skies, the satisfaction of sharing the earth with fellow living creatures, that kicked the gunners into action. Rather, it was the imminent loss of shooting opportunities. When some event — long warned of, and long-anticipated by a few — at last triggers actions by the many, those newly aware masses might deserve credit for finally doing the right thing, but they certainly cannot make claims of having had foresight. Nonetheless, the belated awakening of hunters was the kind of attitude shift that Muir had long hoped for. Late or incomplete epiphanies are better than none.

Still the danger was not universally appreciated, and hunters remained a large and stubborn part of the opposition. When wildlife advocates in Congress finally managed to push through protective legislation on the federal level (the Migratory Bird Treaty Act of 1918, which implemented a treaty signed in 1916 with Great Britain for Canada) they did it against powerful opposition that included midwestern duck hunting clubs. Club members stood to lose their sport if they managed to block the legislation but try they did. Crying states' rights, but at the same time unwilling or unable to take hold of the issue by cooperating with each other on a local level, opponents pushed it all the way to the Supreme Court. In the case of *Missouri vs Holland*, they lost big. As author of the decision, Justice Oliver Wendell Holmes defended the federal interest in migratory wildlife, and summarized the sorry state of affairs that had brought the case to the Court. Note that by "protectors," Holmes was referring to insectivorous birds and their value in controlling agricultural pests—a reason that farm groups had lined up in favor of the law. Holmes wrote:

> "But for the treaty and the statute, there soon might be no birds for any powers to deal with. We see nothing in the Constitution that compels the government to sit by while a food supply is being cut off and the protectors of our forests and crops are destroyed. It is not sufficient to rely on the States. The reliance is vain, and were it otherwise, the question is whether the United States is forbidden to act. We are of the opinion that the treaty and the statute must be upheld."

To be sure, most hunters today agree strongly with the importance of bag limits, defined seasons, and other regulations. But they cannot deny that hunting—irresponsible, unethical, unregulated, and shortsighted—did the damage in the first place. Not all hunters did it, not those who fought hard for sane limits. There were heroes leading the way. But the heroes battled a fierce counter-current of opposition from hunters across the country shouting about states' rights and individual freedom, and denying that problems existed. These were the people who should have been the most aware and the most alarmed, if we are to believe their claims today of being the nation's most effective, pragmatic, and informed of outdoor users. They were not. Not then and, arguably, not now.

For today's hunters to boast that their predecessors moderated their own damage, and to use that as justification for shooting on national wildlife refuges, is like saying that the arsonist who sounded the alarm should be given credit for what the fire fighters managed to save.

"NO ONE KNOWS WILDLIFE LIKE A HUNTER."

In his book, *The Politically Incorrect Guide to Hunting,* author Frank Miniter writes, "Hunters know more about the natural world than any environmentalist I've ever met." To prove it, he describes an unhappy encounter he had with activists protesting a bear hunt in New Jersey. He goes out to see what the fuss is about, and speaks with a woman who is horrified by the killing of bears. She calls it murder. When he acknowledges that he is a hunter and says he also cares for bears, her outrage grows. She can't see how respect for an animal can be compatible with killing it. He tries to explain. She gets more upset, and he gives up. Frustrated by what he feels is her inability to have a calm discussion on the subject, he dismisses her by saying, "she had never even seen a bear. She hadn't spent countless hours under the forest's canopy watching nature, being a part of nature. She was an armchair environmentalist."

His implication: Knowledge of one's quarry, and of nature in general, confers a sort of moral rectitude upon the act of killing the quarry; and that if non-hunters could know the natural world at the same deep level of understanding as a hunter knows it, their feelings would be different.

He's right about one thing. Hunting can open doors of awareness into the natural world. Any time spent outdoors with senses alive and an observant mind results in knowledge. Things happen out there. Insights are gained. First-hand experience is without substitute. But hunting is not the only way into that world of nature-awareness. It is perhaps not the best way. There exist doors to the natural world that a predator, by its predatory nature, cannot open. Predator-sense is profoundly different from quarry-sense. Despite the need for quarry and predator to understand aspects of each other, they remain on opposite sides of a perceptual divide.

Hunting is not a neutral activity in which the hunter can adopt an omniscient point of view, understanding both sides at once (in fact there are many more than two sides). In its best expression — that is, as represented by the most skilled and mindful of hunters — hunting is an attentive way of looking at nature. But it is only one of various ways. The hunting way is biased strongly toward the pursuit of prey. By necessity, the focus of one's senses is tightened. Any house cat watching a sparrow proves the point. The cat's mind is on a narrow track. It perceives subtleties in the sparrow's movements, and vulnerabilities in its behavior that would escape even a devoted ornithologist. That is to say, it has a predator's understanding of the prey-aspects of sparrows. To say that the cat, viewing the world through its predatory eyes, has therefore a full understanding of nature is to ignore the endless complexities — the competing views, as it were — of the living world.

In the same way, a good deer hunter knows a lot about deer and their world, most specifically those things that lead to successful hunting. Yet there are other ways of knowing deer, and of knowing the forest in which they live. Poets, artists, religious contemplatives, scientists, shamans, and philosophers — some of them hunters — have for centuries illuminated multiple pathways of natural understanding.

My point here is that the desire to kill wildlife, regardless of the skills involved, grants no monopoly on understanding animals, not to mention being devoted to their welfare.

Beyond that, intimate knowledge of wild animals and their natural habitat does not in itself justify harming them, any more than understanding dog behavior justifies killing a labrador. The hunter's implication is that with knowledge comes affectionate understanding, which in turn enhances the moral right to take life. This is cloud-cuckoo talk. How can killing something you know intimately hold the high moral ground against not killing that creature in the first place? Knowing one's prey does not change the significance of ending its life. Nor does caring about the creature. These attitudes hold significance for the hunter, for his position in the universe, for his own understanding of the great moral balance between life and death, between responsibility and oblivion. There is a huge difference between the hunter who feels kinship with other creatures and views the taking of wild lives as a sort of communion, and the brute who blasts away in uncaring ignorance. But the difference is to the hunter, not the hunted.

As for the urban bear lover, the least supportable implication of the hunter's argument is that a hunter's killing of an animal he cherishes and respects is morally superior to the city dweller's wanting to protect an animal she has never had the chance to know. A whole barn full of cuckoos make better sense.

Another side to the knowing-your-prey argument

is that knowledge encourages concern for the environment, and actions to protect it.

Miniter asks, "If you gave a group of animal rights activists and a group of hunters a quiz on wildlife, which would get higher marks?" He's talking about knowing the habits of wildlife, recognizing species, perhaps understanding fundamentals of ecology and population dynamics. He is probably right that hunters, on average, have a better command of wild-game facts than the population at large; but he is flat wrong when it comes to non-hunting outdoor recreationists like birdwatchers, plein-air painters, mushroom collectors, backpackers, kayakers, and wolf-watching tourists to Yellowstone National Park. All those millions of field guides, binoculars, spotting scopes, and club memberships represent an awful lot of folks with their attention trained on nature. The millions of hours those people spend watching, learning about, contemplating, and directly experiencing the outdoors add up to a powerful body of expertise.

In any case, the question means little if knowledge does not lead to better treatment of animals and their habitat. We need to ask, does hunting knowledge nurture a conservation ethic? Are hunters more protective of nature than other groups? As it happens, answers do exist. Considerable research has been done on the connection between land stewardship and hunting.

The news for hunters is not particularly good.

In the *Wildlife Society Bulletin,* Robert Holsman of Michigan State University, a hunter and biologist, reviews a host of studies revealing that, in his words, "hunters often hold attitudes and engage in behaviors that are not supportive of broad-based, ecological objectives."

Among the findings cited by Holsman: when it comes to protection of habitat for ecological diversity, many hunters don't get the point. They tend to view wild lands through the lens of what they can shoot; rather like the cat watching the sparrow. Other creatures are tolerated as long as they, or their needs, do not interfere with the production and availability of shootable game. In the eyes of too many hunters, predators are undesirable animals that ought to be shot. They feel the same way about endangered species where management to aid recovery of an animal on the brink of extinction interferes with hunting access. Such a perspective is narrow, dangerous, and ignorant. Holsman concludes that hunters often lack a broad "understanding that producing more deer, ducks, and wild turkeys is not conservation of wildlife, or of natural resources."

Other research draws a correlation between

outdoor use and environmental concern as measured by activities like contributing to conservation groups, attending public meetings on conservation issues, or reading articles and books on conservation policy. In other words, who is more likely to walk the walk? As it turns out, people who engage in non-consumptive wildlife-appreciation activities—whether or not they are also hunters—make better land stewards. Of the outdoor pursuits that foster the conservation ethic, hunting stands at the bottom of the list, a long way below birdwatching.

Even among refuge system employees who hunt, the support for comprehensive environmental issues is strongest among those who also engage in non-consumptive activities.

This is not good news for the refuge system, as it tries to move away from the outdated concept of being duck farms maintaining a supply of huntable waterfowl to a model based more on diversity and natural process. If the science of ecology has taught us anything, it is that true and meaningful conservation in the spirit of Aldo Leopold requires a broad brush. Diversity is an essential aspect of a functioning ecosystem. The wolf and the deer are interdependent, and no species can be dismissed without a loss.

Hunters have a poor record when it comes to supporting non-game animals, especially predators and endangered species. The Recreation Enhancement Act, a funding source to which some hunters (inaccurately) claim to be the major contributors, includes language that prohibits the use of funds for programs relating to endangered species. Senator Larry Craig of Idaho stated at the time of its passage that endangered species research works against the interest of hunters, and therefore hunters should not have to help pay for it. Protection of biodiversity works against the interests of hunters? With attitudes like that, the senator might as well have been standing in a Platte River duck blind, circa 1884, railing against bag limits.

Hunting groups are quick to protect their interests on refuges, even if it means destruction of other refuge animals. Alarmed at the falling population of desert bighorn sheep at Kofa National Wildlife Refuge in Arizona, a coalition of hunting organizations threatened in 2009 to sue the U.S. Fish and Wildlife Service if officials didn't find a way to remove mountain lions blamed for killing sheep, which of course are their natural prey in their natural habitat. Kofa was established specifically for the benefit of desert bighorn, and it could be argued that if lions endanger the sheep population, then action might be warranted. There has been disagreement even among wildlife professionals on whether the lions present a threat to bighorns. On the other hand, the hunting organizations, in saying they wanted to protect sheep, did not offer to stop killing the sheep themselves. It's the old story: "Kill the predator, because we want to kill the prey ourselves." In 2010, Arizona issued six sheep tags on or immediately adjacent to the refuge; meanwhile, the refuge adopted a policy allowing the killing of lions that kill sheep.

Another painful example of this can be found in the ongoing fight over lead in bullets, buckshot, and fishing weights. Recently, a coalition of hunting groups including the Congressional Sportsmen's Caucus (whose members are United States senators and representatives) lined up to protect the use of lead despite solid evidence of environmental damage. The damage is well enough proven that 20 years ago the U.S. Fish and Wildlife Service issued a nationwide ban on lead shot for waterfowl hunting. The ban was to protect water quality and to prevent birds from eating spent shot while feeding, and being poisoned as a result. It was a good move, proved by the fact that 20 years later, trumpeter swans and other waterfowl are still dying from ingesting lead that was scattered over wetlands, including refuge lands, before the ban went into effect. Meanwhile, lead shot remains generally legal for hunting dry-land birds like pheasants, turkeys, and doves. Lead rifle bullets are banned only in limited sensitive areas, such as condor territory in California. Consequently, wolves, foxes, bears, eagles, and other scavengers swallow lead during hunting season when they feed on gut piles riddled with metal from bullets designed to fragment when they hit. Some die as a result.

It's not only animals at risk. Lead-tainted meat in game birds has long been recognized as a hazard to humans. It gets into birds directly when hunters shoot them with lead pellets. Being small, the pellets can be hard to find. Some remain in the meat and make it all the way to the dinner table. Lead also gets inside birds when they pick up spent pellets lying on the ground. Millions of pounds of lead pellets are spilled every year by dove hunters alone. Densities of more than 350,000 pellets per acre have been recorded on popular dove-shooting areas. Birds need pebbles for their gizzards, and pick up lead shot along with harmless sand and bits of gravel. Two or three pellets in a gizzard are enough to kill a bird. Worse yet, if that bird is shot before it dies from lead poisoning, the meat can be hazardous to humans who have no way of knowing it is tainted.

The same is true for deer, elk, and other large animals. Game meat has been found contaminated with lead at levels far beyond safety standards for human food. The North Dakota Department of Health, in 2008, issued a state-wide alert after X-rays revealed lead fragments in 53 of 95 tested packages of ground venison. The same year, the Wisconsin Department of Health and Family Services issued warnings that lead shards too small to be detected during normal processing were found far from the bullet path in a significant portion of samples tested. Wisconsin advised charity food pantries not to distribute donated venison unless it had been cleared by X-ray analysis. What began as natural, free-ranging wild meat ended up toxic. Hunters are the primary consumers of questionable meat, yet they continue to resist replacement of lead with non-poisonous ammunition. Lead-free ammunition costs more but is readily available.

Why is this important to national wildlife refuges? Most refuges already require non-toxic shot for all bird hunting, not just waterfowl. But lead slugs and bullets are permitted on refuges for big game, and few restrictions exist on lead used by anglers. Lead is perhaps the only toxin that can be applied legally to refuge lands, and it's not nature photographers doing it.

Beyond that, the question of stewardship is central to the way refuges are managed, and it's not entirely up to the managers to decide what's best. Through organizations like the National Rifle Association, hunters wield disproportionate political clout. Politicians are almost required to demonstrate hunting credentials. Holding a rifle is as much a political cliché as holding a baby. As the history of refuges proves, what hunters demand, they often get. Their demands can be beneficial, as in supporting the existence of refuges in the first place; but damaging if they press for policies that distort the primary mission of the system. Killing mountain lions to make more sheep for hunters to kill is an example of hunter-centric not wildlife-first policy.

With all due respect for mindful hunters who understand the natural world beyond the narrow periphery of their sport, it would help wildlife if such hunters were not so rare.

"WE BOUGHT THEM."

We hear from all directions that hunters foot the bill for wildlife through Duck Stamps, excise taxes, and memberships in organizations like Ducks Unlimited, the Safari Club, and the National Rifle Association; and that non-hunters who appreciate the sight of wild animals get a free ride thanks to the generosity of hunters. On the U.S.F.W.S. website: "The sale of hunting licenses, tags, and stamps is the primary source of funding for most state wildlife conservation efforts." In a 2001 report for the National Center for Policy Analysis: "The money hunters spend and contribute pays the cost of wildlife protection." The Congressional Sportsmen's Caucus, comprised of U.S. senators and representatives, referring to hunters and anglers: "The best stewards of our environment, as well as the financial backbone to fish and wildlife conservation in our country."

It's simply not so. Not for the refuge system. Not for the country as a whole.

In fact, hunters provide only a fraction of the refuge system's funding. The annual budget, appropriated by Congress from general tax revenues, comes to around $500 million. Hunters account for less than 5 percent of the U.S. population. As such, they provide a proportional share of those revenues through general taxes, like everyone else in the country. That share is about 5 percent if we assume that hunters, on average, pay average income taxes. In other words, an average hunter pays no more and no less than any other average taxpayer.

As for land acquisition, it is true that Duck Stamps have paid for an important but small portion of the system's 150 million-plus acres. Their main achievement is the establishment of some 26,000 Waterfowl Production Areas, mostly in the north-central states. These are critically important places, and credit must be given. They provide essential habitat for migratory waterfowl, and they were paid for almost entirely by revenues from Duck Stamps, to the tune of more than $700 million dollars and counting.

The other 96% of refuge land—from 3-acre Pelican Island to the 19 million acres of Arctic

National Wildlife Refuge—was carved out of existing public holdings, or paid for by programs that have no direct connection with hunting.

Red Rock Lakes National Wildlife Refuge, in southwestern Montana, is a good example of support coming from many different directions. Established in 1934 to save America's last known nesting population of trumpeter swans, it measures 47,756 acres. Of that, 9,812 were acquired by withdrawal from the public domain. Another 22,628 acres were bought during the Great Depression by the Resettlement Administration (later part of the Farm Security Administration) using federal funds for the relocation of struggling rural families. In later years, Land and Water Conservation Act funds paid for 4,936 acres, and 9,355 acres were added from other sources including gifts from private donors. Duck Stamp revenues provided 1,024 acres.

When hunters speak of their financial contributions, they are usually referring to money generated from the sale of state hunting licenses and taxes levied under the 1937 Pittman-Robertson Act (officially the Federal Aid in Wildlife Restoration Act). It's important to note that money from license sales plays no part in funding the federal refuge system. Neither do Pittman-Robertson funds. Their purpose is to support *state* wildlife agencies through excise taxes on guns, ammunition, bows, arrows, and other hunting equipment. Anglers point to a similar law, the Federal Aid in Sport Fish Restoration Act (also called Dingell-Johnson) which applies to fishing gear. These funds are substantial in scale. Since inception, Pittman-Robertson has raised nearly $6 billion.

It's hard to imagine what state agencies would do without those dollars. On the other hand, critics point out that the main use of hunter-generated money is to manage hunters, hunting, and huntable wildlife at the expense of other animals. There is an element of truth in this; recall, for example, the sentiments of Senator Craig. Early wildlife management was essentially a matter of trying to control hunters. Aldo Leopold himself, on the first page of his classic book *Wildlife Management,* defined the discipline as "the art of making land produce sustained annual crops of wild game for recreational use." In other words, the prime object of wildlife management was to provide hunters with a never-ending supply of healthy targets.

More recently, historian Robert L. Fischman, in his book *The National Wildlife Refuges*, writes that the 1997 refuge law "requires the use of 'modern scientific resource programs,'" and suggests that this means management for broad ecological health of a refuge rather than old-fashioned encouragement of game species for the benefit of hunters. He adds, "But most wildlife management in the United States today *is* oriented toward hunting concerns…. And, many problems with game management center on the narrow measures of success (e.g., single species populations) more than the actual practices employed (e.g., habitat enhancement).

For many years, a big chunk of state wildlife-management money went back out in the form of bounty payments on so-called "noxious" species like raccoons, crows, bobcats, coyotes, weasels, raptors, and others that hunters regarded as competitors for the things hunters wanted to shoot. Alaska, that bastion of free-thinking hunters, once offered a bounty on bald eagles — a state subsidy for killing our national symbol.

Things improved significantly after President Nixon signed the Endangered Species Act in 1973. The law required state agencies to broaden their scope to promote ecological diversity. Yet it remains an important issue, this matter of how hunter-generated money affects wildlife management priorities. It raises broad questions about stewardship and the overall health of our environment. Where hunters are the paymasters, they exert a de facto pressure on managers who draw up the policies. However, hunters are not the paymasters of the refuge system. Pittman-Robertson taxes and license fees, whatever their benefit to wildlife might be, contribute nothing at all to national wildlife refuges. Except, perhaps, to give hunters a sense of entitlement.

On the subject of entitlement, hunters argue that birdwatchers and other "non-consumptive" users are getting a free ride, enjoying wildlife without paying their share of support for conservation programs. Proposals are floated periodically to initiate a tax on things like hiking boots and backpacks similar to the excise tax on guns and ammo; or to require somehow that non-hunters buy Duck Stamps. (An unknown number do just that, as a good-faith contribution to the waterfowl everyone loves to see, and because the stamps are beautiful, collectible things in their own right.)

The idea is that all users should contribute their share, because all users share the benefit. This is certainly valid. On the other hand, hunters get a larger share than other users. If successful, they go home with meat. They take the ducks out of the sky, and the moose out of the marsh. They also cost more to manage, and history has proven the ongoing necessity of hunter management to prevent poaching and other abuses. Birdwatchers, hikers, and photographers need far less supervising. After all, doing criminal damage to swans with a spotting scope is not easy.

It seems only right that hunters should pay for their consumptive actions. Most do so willingly, even proudly—while still making claims of being unfairly burdened for their way of loving and appreciating nature. How would a non-hunter answer that? With a simple suggestion: Hunters can easily put things on an equal footing with photographers and birders. They can join them. They can go out with binoculars and bird books. They have this right any time, no license required. And, even better, they don't have to give up hunting. They can hunt any time of the year for any species in virtually any place including national parks. Like anyone else, a hunter is free to prowl the woods or suburbs with, let's say, a walking stick, point it at a wild creature, and pull a pretend trigger. In doing so, he would get all the much-touted benefits of hunting (short of the meat) and pay no more than a hiker. The hunter would have his close relationship with nature, his time in the outdoors, and his chance to pursue quarry. He'd still get to walk quietly, learn all the necessary woodcraft, do it in the company of close comrades, and tell stories around the campfire—in short, all those things hunters cite in defense of hunting. The only thing missing would be the carcass.

That may seem an unlikely—perhaps ridiculous—suggestion, but it's not. In a deer-hunting parallel to catch-and-release fishing, the American Whitetail Authority runs an annual competition in which hunters prove their skills using rifles loaded with blanks. Digital recorders in their telescopic sights document the "kill," provide electronic trophies, and show the quarry bounding off to be shot another day. Some of the most skilled hunters in the country turn out for the event. They win prizes, and earn boasting rights. The only thing missing is the carcass.

❧

In the end, the question of who supports wildlife in America is larger and more encompassing than any interest group. The refuge system, although enormous, is only a sample of the vast ecological web of life. All wild lands, all habitat, private or public, in America, Canada, Mexico, and other continents—even the international territory of the open seas—are bound to each other by a network of fine ecological strands. When we speak of conservation, the actions of all governments and organizations and individuals have an impact, especially those whose mission is to protect wildlife and natural habitat. Hunting organizations, the likes of Ducks Unlimited, Pheasants Forever, and the Boone and Crockett Club, do play a role, sometimes an important and beneficial one. But their efforts are joined, and overshadowed, by the larger universe of concerned entities and individuals who do not hunt on refuges, or for whom hunting is not a central issue. That universe includes the national parks, national forests, state parks, community arboretums, school forests, farm woodlots, backyard wildlife gardens and any number of other entities, large and small. These include The Nature Conservancy, The International Crane Foundation, the World Wildlife Fund, the Wildlife Conservation Society, the Humane Society's Wildlife Land Trust—even Walmart, which runs a program called Acres for America. Through that program, Walmart shoppers buying everything from detergent to light bulbs have funded the protection of more than 625,000 acres for wildlife.

The National Park Service budget for 2010 was more than $2.25 billion. Although a share of that total went to urban parks and historic parks whose mission is not directly wildlife-related, the overwhelming majority supports areas critical to wildlife—Yellowstone, Grand Canyon, Great Smokies, Denali, Olympic, Glacier, and so many others. The money that keeps national parks healthy comes almost entirely from general funds, allocated every year, and we have all Americans to thank for it.

In the November, 2010 elections, voters across the country approved new state and local taxes, bond measures, and other levies totaling almost $2.4 billion in support of conservation, parks, open space, and wildlife habitat. Since 1988, similar measures have approved a whopping $56.3 billion for conservation. These were not taxes on a specific user group. They are but recent examples in the long history of the general public supporting conservation in their states and local areas.

As for wildlife research, much of the work done by state fish and game agencies is supported by hunter dollars, but an enormous contribution is also made by scientists from hundreds of other organizations, agencies, universities, and government bureaus who get their money from a wide variety of sources, both public and private. Putting a number on all those efforts is difficult, but surely it surpasses the research income from hunters.

This is not to trivialize the substantial contributions of hunters and the shooting sports industry. These are matters where every contribution is needed. But shooters are only one part of a very large picture. Hunters should stop saying they do more for conservation than anyone else. It makes them sound small and uninformed.

REFUGE SYSTEM FUNDING

Congressional appropriations pay for operations and maintenance as part of the federal budget. This is the major funding each year, about $500 million. Hunter contribution to refuges: proportional to population.

Duck Stamps (Federal Migratory Bird Hunting and Conservation Stamp), administered by the Migratory Bird Conservation Commission, are required of all waterfowl hunters but available to anyone. The stamps also serve as entrance permits at refuges that charge for admission. Funds are used by the refuge system to buy or lease wetland habitat, primarily in the north-central states. In recent years, Duck Stamps have earned about $25 million annually. Hunter contribution to refuges: almost all of this category.

Excise taxes on firearms, ammunition, bows, and arrows (Pittman-Robertson Wildlife Restoration Act,) and fishing equipment (Dingell-Johnson Sport Fishing Restoration Act) go directly to the states where it is matched at a 25 to 75 percent state-to-federal ratio, and used primarily to support the work of state fish and game agencies. Some states contribute to federal Wildlife Management Areas, part of the refuge system, but there is no requirement that any funds be spent for that purpose. Hunter contribution to refuge system: minimal.

The Land and Water Conservation Fund gets money from sales of surplus Federal property, taxes on motorboat fuel, and oil and gas leases on the Outer Continental Shelf. In recent years, almost 100% has come from oil and gas leases. The fund is intended primarily for acquisition of land for recreational purposes, not specifically for refuges. The President proposes how to use it, as an item in the budget subject to Congressional appropriation. Congress does not always appropriate these funds, and a variable amount ends up in refuges. Hunter contribution to the refuge system: proportional to population.

The Recreation Enhancement Act, passed in 2004, allows refuges to collect entrance and user fees. The proceeds support recreational uses. In 2009, the U.S. Fish and Wildlife Service took in $4.7 million. If hunters contributed 25% of the total (an estimate based on hunter-to-nonhunter visitor ratios) their contribution to refuges: about $1.3 million.

"IT'S A MANAGEMENT TOOL."

Management of wildlife by hunting always means reduction. You don't hunt something to make more of them. Besides that, it almost always refers to a few species—notably white-tailed deer and Canada geese—that become a nuisance by destroying crops, gardens, and landscaping, or in the case of geese, covering golf courses and city parks with squishy droppings.

On refuges, however, management should mean something else. Refuge management is not a question of keeping the rose bushes looking good, or having clean grass for picnics. Rather, it's about the health of the natural system. Deer are a good example because they are resilient. They come back after hard times, and respond exuberantly where food is abundant.

Reduction by hunting yields only a short-term result. But swarms of deer can be hard on refuge habitat, on other species trying to survive alongside them, and on neighboring lands.

Wildlife managers say hunting is the most effective tool they have to control deer populations. At the same time, they acknowledge that hunting by itself is not enough. In fact, it's getting less effective. The reasons for this are complicated.

They include the decreasing popularity of hunting, meaning that managers can't rely on having enough hunters show up to meet reduction objectives. Also, hunters want to take bucks for their trophy value while managers want them to go after does. Shooting bucks has a smaller impact on the population but hunters balk at regulations that encourage or require them to shoot does. Another factor is the increase of areas where hunting is not allowed at any time. These include rapidly expanding suburban and semi-rural areas where deer make out very well living among widely spaced houses on nicely landscaped lots, but can't be shot for obvious safety reasons. Added to that, increasing numbers of private landowners forbid hunting for a variety of reasons, many of which have nothing to do with game management.

The result is predictable. If safe zones exist, the deer will find them. If there's something to eat, the deer will stay. Increasing the hunting pressure outside safe zones is not much use. You end up with more deer inside, and the problem gets worse.

For refuges to become lifeboats overcrowded with deer or any other species is clearly not a good thing.

On the other hand, most species do not threaten to overload a refuge, and for them, the hunting question has different answers. Controlling overpopulation is not the same as taking animals because you can — that is, because the population can withstand the loss of some individuals. Another way of asking the question: what is the reason for hunting if the refuge doesn't need it?

In many cases, as a matter of general policy, refuge managers are told to cooperate with state fish and game agencies. The refuges can decide for themselves whether to do that. They often apply more restrictive rules, or prohibit taking species that are of particular interest to a given refuge. But it's common for a refuge to simply fit into the stream of state agency decisions, whether or not this is to the advantage of the refuge or its wildlife.

A case in point is the proposal by Red Rocks Lakes to expand elk hunting to areas that have been traditionally closed. This meets no refuge objective. Elk are not overgrazing the place. In fact, they gather on the refuge during hunting season in search of safety.

The refuge is a wide open place, where elk can easily be seen, and easily hunted. If hunting is allowed in these open fields, the elk will learn not to come there. They will stick to areas with better cover. Hunters will still have access to them, albeit with more difficulty, but other visitors will lose a splendid viewing opportunity, and the refuge will gain little if anything in terms of its own management goals.

So why allow it at all? Refuge managers themselves have their doubts, according to a widely distributed 1989 GAO report on compatibility of refuge activities. Refuge managers across the country were questioned for their views on activities ranging from mining to water skiing to hunting. The report stated, "Refuge managers did not believe that all uses they regarded as harmful should be discontinued on their particular refuges. Managers we talked to said they were sometimes willing to accept the adverse effects of some harmful activities as the price of obtaining the good will of the local public or various economic interests. In about half of the instances, though, refuge managers reported that the price was too high and believed the harmful secondary uses should be discontinued. When viewed as harmful, military air and ground exercises, mining, logging, and *waterfowl hunting* were cited by the highest percentage of managers as meriting discontinuance." [italics added]

They might not be able to, whatever they believe. As one manager put it, "There's a difference between political carrying capacity and biological carrying capacity." He was referring to the way political pressure interferes with scientifically based management decisions. The pressure usually works in favor of increased hunting.

Discussions of hunting quickly edge into the concept of so-called "surplus" populations, and Aldo Leopold's definition of wildlife management as "the art of making land produce sustained annual crops of wild game for recreational use." A crop is meant for harvest. You take the surplus, and leave the productive base.

A surplus, however, is in the eye of the beholder. To a hunter, it means the number of animals that can be taken without putting the general population at risk; it views wildlife as a renewable commodity that can be harvested at controlled levels. Hunters are all for abundance, but they also place a contradictory premium on scarcity, which makes an individual animal special and therefore more desirable; they also value wariness on the part of the quarry because it makes hunting more challenging and bestows additional credit to the skillful hunter.

On the other hand, to a wildlife watcher, surplus means a larger herd, a bigger flock, more chances to see a critter. Limited only by the habitat's ability to sustain a healthy population, the more the better.

Wildlife watchers are delighted when large animals pile onto the refuges looking for safety, while the hunters who chase them there, or who find animals scarce elsewhere, are frustrated. Hunters want access. They lobby hard for it, claiming that those animals are rightfully theirs to take. So far, the trend has been in the hunters' favor.

Critics of hunting on refuges make further claims: that alternate methods exist for population control; that the need for control is overstated as an excuse for shooting; that hunting is a blunt instrument that can backfire. Hunters don't always behave as managers would want them to (for example, shooting legally at tundra swans but hitting the occasional endangered trumpeter; or the hunter's tendency to prefer the healthiest, most viable of animals, rather than the smaller, older, or weaker).

Population dynamics in wildlife is an enormously complicated subject—too large for this small book. In the end, the important points about hunting as a management tool are these: 1) hunting is not very effective, and is becoming less so; 2) it can be a blunt tool that causes collateral damage; 3) it can become a political football in a game where the rules do not favor wildlife. As a justification for killing animals on a refuge, it holds not much air.

A GREATER GOOD

With the National Wildlife Refuge Improvement Act of 1997, Congress declared that hunting on refuges is compatible with their purpose.

Congress declared it so but set no goals, nor specified any targets for how much hunting should be allowed, except to say that wildlife-dependent recreation should be encouraged. On the other hand, refuge managers have been given powerful tools to limit recreational access, and could do so with the right scientific and political backing. They could limit hunting. They could limit birdwatching too.

Here we get into an area that seems crucial, but lies in the gray zone of management decisions. It's the question of benefit, and goes like this: if wildlife come first, and refuges are closed until opened, shouldn't any use have to prove itself to be of benefit to wildlife, or not be allowed? The notion has been supported by numerous experts, managers, and testifiers over the decades, that by the highest standard for compatible human use, every action should be beneficial or at the very least, neutral. Bird watchers or bird shooters, it shouldn't matter. If an activity harms wildlife, it should not be allowed.

But what is harmful, and whose definition should prevail?

A few pertinent facts: Recreational use of refuges increased more than 80% from 1985 to 2005. Most of the increase is non-lethal use. The number of hunters is falling nationwide, while the number of non-hunting refuge visitors is fast rising. The majority of Americans support hunting for food or population management, but sentiment runs the other way about hunting for fun or trophies. The national trend is away from consumptive use of refuges and wild areas in general, and it's a trend that wildlife professionals themselves reflect. Whereas agencies like the Fish and Wildlife Service were once staffed primarily by hunters and anglers, their numbers are falling, and attitudes are changing. In a 1998 survey of wildlife professionals, "only 52.5% agreed with the statement that 'Wildlife and fish species are resources to be harvested in a sustainable way and used for human benefit.'" That's an astounding figure, given the history of wildlife management. It bears repeating. Only a slim majority of wildlife professionals believed that the purpose of wildlife was to provide a harvestable resource. (Harvest, an unfortunate and misleading euphemism that ought to be abandoned, is their word, not mine.)

University courses in wildlife management will surely follow the same trend. As fewer students with hunting backgrounds enter the field, academic courses will move away from their old emphasis on what the profession calls "species harvest management" toward a broader, more inclusive view of wildlife appreciation.

The historical trend is clear. If the refuge system reflected the trend, management would be moving away from hunting being the so-often dominant use of refuges, and away from refuges being used as wildlife farms to supply hunters with prey. In fact, the latter is a stated goal. But then why is the system moving at such speed toward maximum expansion of hunting? They say it themselves, the refuge managers, the politicians, the shooting sports industry — they warn that hunting is becoming less popular and that hunters need encouragement and nurturing or hunting could become an historic curiosity. It sounds a lot like the language early conservationists used in reference to ducks when waterfowl populations were at crisis levels. Something had to be done to save the ducks. Now it seems we have to save hunting. Policy makers act as if there is some invisible-ink hidden language in documents establishing refuges that says "for the benefit of hunters, and the preservation of hunting." Whether or not the words are there, hunting supporters constantly echo that sentiment. As justification, they draw a direct line between hunters and the welfare of the animals they hunt. That line, as it exists today, is not so direct, especially in the case of national wildlife refuges, where hunters take far more than they give, despite a growing need of support from the many other wildlife-interested constituencies.

ꞔ

Despite its enormous lobbying power, hunting suffers from a public relations deficit.

The hunter of legend is an iconic figure, upright, silent, and strong, who glides with grace and skill through the romantic literature of the American frontier. Natty Bumppo, Uncas, and Chingachgook join real-life kings of the wild frontier like Daniel Boone, Jim Bridger, and Davy Crockett to conjure an idealized image of the lone hunter. The image remains strong today. A stalwart fellow drifting shadow-like through the forest, his feet touching ground like falling leaves, all senses alive, attuned to everything around him, not just his prey but the entire nuanced web of life in the wild. He takes his place as a native son, a welcomed participant in the communion of life. When he shoots, he kills with a clean respectful shot. His is a personal relationship with nature that by its nature remains private, discreet, and even sacred. The image borrows from Native Americans, for whom taking of life was—and for many still is—an acceptance of nature's gift of sustenance. In some beliefs, the gift is a sacrifice by willing quarry to the deserving hunter, whose own spirit must be aligned and respectful, or he will fail.

There are hunters today who think and act with reverence of that sort. Yet the public face of hunting is nothing like it. In waterfowl marshes, shotgun barrages shatter the quiet of dawn. In forests and cornfields, rapid shots echo from semi-automatic weapons or from groups of hunters blasting at distant, fleeing, or indistinct targets. Jacked-up pickup trucks, horse trailers, ATVs, and fresh carcasses are the modern face of hunting. Irresponsible shooters abound, and be it fair or not, they are a reason why many non-hunters are afraid to take a walk in the autumn woods.

Inevitably, there are the trophies. Hunters pose with rifles and dead bodies, in postures that communicate not love for the vanquished, but rather pride in the doing. Mounted heads and feet and whole animals hang on walls and seem, in the eyes of many non-hunters, to represent anything but expressions of love and one-ness with the natural world.

Their conservation credibility is too often conflated with gun-rights issues; their agenda too easily hijacked by powers like the Congressional Sportsmen's Caucus, which, as the conservation writer Ted Williams and others, including the League of Conservation Voters, have pointed out, is populated by enemies of the things hunters profess to hold so dear.

It does not have to be this way. The business of protecting wildlife is a much too complicated issue to be seen in simple terms. Hunters could be, and should be, natural allies with all conservationists — even those who oppose hunting. Instead, they are prying open the doors to shooting in state parks and other previously closed reserves, and pressing for expansion of permitted seasons, species, and territories on wildlife refuges. Most of their fellow citizens do not support this. Hunters are right to be looking over their shoulders, defensively, at the rising swell of critical opinion.

ↂ

Americans' relationship to wild lands and wildlife continues to evolve. Throughout American history, wild animals have nourished us both physically and emotionally, providing food for the body and respite for the spirit. Refuges have played an important role in protecting and sustaining wildlife. Their importance can only increase. Meanwhile, hunters have become minority users, while millions more crave the sort of contact with nature that refuges provide. John Muir, so often prescient, may have understood a century ago what the majority of Americans would most need from wild land when he spoke of Roosevelt's hunting. "Mr. Roosevelt," he said to the president, "when are you going to get beyond the boyishness of killing things.... Are you not getting far enough along to leave that off?" Roosevelt famously answered, "Muir, I guess you are right."

Although Roosevelt never stopped hunting, he believed that wildlife should have inviolate sanctuaries. In his view, refuges would serve as nurseries and storehouses to hedge against destruction occurring elsewhere. To be sure, he expected that the animals nurtured in refuges would move outside their boundaries and become available for hunters. He never envisioned that sanctuaries themselves would be opened for hunting, and it seems very unlikely that he could have foreseen the Faustian bargain that would arise from a marriage of refuges and hunting.

It's a simple equation: If refuges are well-managed and attractive to animals (they aren't always attractive, but these are questions of management beyond the scope of this book) they become purveyors of that bargain. Wildlife is drawn to them as songbirds are to a feeder, or as thirsty moose are to a garden pond. And there, they die. They accept the food, the shelter, the protected nesting ground, without knowing that these attractions have been provided—in the eyes of some, including refuge administrators—to make them available for killing.

The attitude is reflected by the ubiquitous use of the euphemism, harvest. You don't kill an elk, you harvest it. As if it were an ear of corn. As if the hunter had planted a crop, and was collecting the fruit. The word carries implications of earned reward, of husbandry and harmlessness, and mutual benefit. It's a faulty analogy. Elk aren't wild raspberries, asking to be taken. Designed by nature as a means of propagation, berries are offered, bright and tasty, to birds and other animals as a way of getting their seeds dispersed. Grasses use a similar tactic. They brown off in the fall, and present their mature grains to grazing animals whose hooves churn up the soil and plant the seeds of next year's growth. From the consumption of seeds and fruits, there flows a mutual benefit designed by nature.

It doesn't work that way for the pregnant cow moose who leads her calf onto refuge grounds in September, seeking forage and protection only to become the target of a hunter's bullet. She benefits not at all from dying. Her embryo, that seed of the future, dies with her. Her orphaned calf will be lucky to make it through the winter. One bullet kills three.

Hunting groups might point out that the moose, by being shot, has proved her value to humans, and therefore she has won political and economic support for refuges, which in turn provide habitat for other moose and for wildlife in general. And that by shooting the moose, hunters reduce moose numbers and assure an adequate supply of forage for the animals left standing. These arguments are partly true. Yet they only underline the nature of the relationship.

In his 1972 novel *Watership Down,* author Richard Adams described a chilling scene. The story is told from the perspective of wild rabbits forced to migrate when their warren is destroyed by a development project. They undertake a journey with many hazards and adventures. Along the way, they meet a colony of rabbits, smug and well-fed, who invite them to settle down. There's room for more. Life is good. The neighboring farmer tosses out quantities of good food, and shoots the predators. No wonder these rabbits are so big and healthy, yet why does the warren contain so many empty burrows?

To the wanderers, it feels like paradise. They want to stay, until they learn, to their horror, of the bargain.

The colony accepts that the farmer will snare a portion of their members for his pot. They accept this in return for the generous supply of food and protection from foxes. The survivors have grown fat. They've been blinded by comfort, to the point that they ignore their peril and the loss of their comrades. The wandering rabbits discover the truth and flee for their lives.

Although its main characters are wild rabbits, *Watership Down* is not really about wildlife. This story of the fat rabbits is a parable, directed at a human audience, about the dangers and hidden costs of being too comfortable. Nor is it critical of the farmer, who is simply taking advantage of an easy source of meat. It certainly does not suggest that wildlife is corrupted by protection or the easy availability of food, as if cranes coming in to a cornfield have calculated and accepted that the price of corn is the risk of death. Yet it does illustrate view, widely held among hunters, of the human relationship to the natural world. According to this view, hunters play the role of the farmer, even to the point of their adopting agrarian terminology. They call it harvesting, but it's not. Laying out food bait, or enhancing habitat to attract wildlife so you can kill it is not the same as planting a crop; and it doesn't imply some sort of entitlement to take the lives of the animals that present themselves as a result, although many hunters seem to think it does. They are wrong.

A dead crane was not part of any bargain. It wasn't harvested. It was killed.

There is no doubt that hunters have contributed to the protection of land and wildlife. But so often they express their support as if they've negotiated the terms of a one-sided deal: "We will feed you and shelter you, and protect your life, until we decide to take it." In this lies the suggestion, sometimes posed as a threat, that without the opportunity to take their share, support will be withdrawn. If there's not hunting, we are told, wildlife will suffer.

If that seems wrong—that hunting is necessary to motivate some people to vigorously support wildlife conservation—it should nonetheless energize non-hunters who care about the survival of the wild. It goes doubly for those who oppose hunting on wildlife refuges, or want to see it reduced rather than expanded. Opposition by itself will never be enough. Wildlife needs all of us.

❧

Gifford Pinchot, the first director of what would become our system of national forests, is widely known for his philosophy on land and wildlife management. "Where conflicting interests must be reconciled," he wrote, speaking of wild lands in general, "the question shall always be answered from the standpoint of the greatest good of the greatest number in the long run." By *number*, he meant the number of people, who might be counted according to a particular interest. By *good*, he meant a value—something not always quantifiable and certainly subject to interpretation.

Pinchot's contemporary, John Muir, found the concept coldly utilitarian. In fact, the word utilitarian came from British jurist and social reformer Jeremy Bentham's philosophy of Utilitarianism, reinforced by John Stewart Mill in his 1863 book *Utilitarianism*. Pinchot was a believer who felt that the greatest good could be measured, quantified, and therefore recognized among a slate of choices. His bias was toward action—doing things, using things, rather than leaving them alone. "The first great fact about conservation," he said in 1913, "is that it stands for development. There has been a fundamental misconception that conservation means nothing but the husbanding of resources for future generations."

Muir, as always, had his eyes on higher ideals and longer horizons that could not be set so neatly on a balance. The battle over damming Muir's beloved Hetch Hetchy Valley in Yosemite National Park was a clear example of the differences between the two men. The impounded water would serve a great city—San Francisco—and Pinchot saw that as the greatest good. As he expressed it in 1910, "If we had nothing else to consider than the delight of the few men and women who would yearly go to Hetch Hetchy Valley, then it should be left in its natural condition. But the considerations on the other side of the question, to my mind, are simply overwhelming.... I never understood Muir's position on Hetch Hetchy."

Muir saw a greater good (but one more difficult to measure) in the preservation of an irreplaceable natural treasure. "The Phelans, Pinchots and their hirelings," he declared, "will not thrive forever.... These temple destroyers, devotees of ravaging commercialism, seem to have a perfect contempt for Nature, and instead of lifting their eyes to the God of the mountains, lift them to the Almighty Dollar. Dam Hetch Hetchy! As well dam for water-tanks the people's cathedrals and churches, for no holier temple has ever been consecrated by the heart of man."

Pinchot, the man of action, prevailed. Hetch Hetchy drowned.

Years later, Aldo Leopold, who had been a student of Pinchot at Yale, and an admirer, found Pinchot's concept incomplete. He felt that judging an action by its effect on people alone would inevitably lead to the deterioration of the natural system. Not only human needs, but also those of nature had to be included in the concept. To consider the land was essential.

Otherwise, it could not be sustained.

Bentham himself might have agreed. In addition to Utilitarianism, he held strong views on the rights of living things. He wrote in 1823 to condemn slavery and the laws by which some people were put on the same footing as "inferior races of animals." He then predicted that some day not only humans but "the rest of the animal creation may acquire those rights which never could have been witholden from them but by the hand of tyranny." He scorned the use, prevalent among other thinkers at the time, of reasoning ability as a means of judging who should or should not have rights, saying that a "full-grown horse or dog, is beyond comparison a more rational, as well as a more conversable animal, than an infant of a day or a week," and then delivered his own standard: "The question is not, *Can they reason?* nor, *Can they talk?* but, *Can they suffer?*

To think that animals might have rights? Some cultures have felt so forever, long before Bentham made his prediction. In the modern world it remains a persistent, if unpopular, notion.

A conversation among the four men — Bentham, Muir, Pinchot, and Leopold — is one we can only imagine. Yet nearly 200 years after Bentham coined the phrase about the greatest good, living as we do under the grinding and increasing pressures of the 21st

Century, at a time when non-lethal users of wildlife refuges represent a dominating majority, with wildlife threatened from all sides (it must be recognized that regulated sport hunting ranks as a minor threat compared to other destructive forces here and abroad), the two once-disparate standards seem less far apart. In 1913, Pinchot thought San Francisco needed Hetch Hetchy's water more than the nation needed Hetch Hetchy. Is that still true? Will it be true in another century? Muir, for his part, would have been appalled to know that wildlife refuges would some day be opened for hunting. He would have spoken against it in high-minded, non-utilitarian terms. But times have changed. Today, although he might see it as lower-level thinking, he could pragmatically adopt Pinchot's standards. He could use population numbers to defend Hetch Hetchy Valley. He would find that in the past century, the greatest number has come around to his sense of how to appreciate nature—that is, to agree with his concept of the greater good. Considering wildlife, he might well borrow even Pinchot's language and ask "If hunting on wildlife refuges does not serve the greatest number, then how can expanded hunting on refuges be the greatest good?"

In 1968, Aldo Leopold's son A. Starker Leopold, a renowned scientist in his own right, chaired a blue-ribbon advisory committee on wildlife management on the refuges. In their report, the committee acknowledged the sometimes conflicting views on the role of refuges, and attempted to set a broad philosophical objective.

The committee wrote, "We view each National Wildlife Refuge in the old-fashioned sense of a bit of natural landscape where the full spectrum of native wildlife may find food, shelter, protection and a home. It should be a place where the outdoor public can come to see wild birds and mammals in a variety and abundance compatible with the refuge environment. It should be a 'wildlife display' in the most comprehensive sense."

They were saying that refuges should represent slices of America's natural wildlife diversity. They were speaking up for things not hunted, for non-game animals, and for Americans who wanted to experience that full-panoply display. "The number of Americans concerned with viewing or photographing wildlife," they wrote, "is increasing at least exponentially with population. Their interest should be served by the refuges, along with the interests of the hunting public."

Hunting was an acknowledged use of refuges, but they asked a question that shows their awareness of how things could change. It's the same question Muir might have posed about greater good, but in different words. "In America of the future," questioned the committee, "what are likely to be the highest social values that the refuges can serve?"

Half a century later, the question of killing animals on wildlife refuges—where, how many, which ones, and whether to do it at all—deserves a fresh look.

Books by Jeremy Schmidt

*Himalayan Passage: Seven Months in the High Country of
Tibet, Nepal, China, India and Pakistan*

Grand Canyon, The Life and Times of a National Treasure

The Saga of Lewis and Clark: Into the Uncharted West
(with Thomas Schmidt)

Grand Teton: Citadels of Stone
(with Thomas Schmidt)

In the Village of the Elephants
(with Ted Wood)

*Two Lands, One Heart: An American Boy's Journey
to His Mother's Vietnam*
(with Ted Wood)

Grand Canyon: A Traveler's Guide

The Rockies: Backbone of a Continent

Adventuring in the Rockies

Dolls and Toys of Native America

In the Spirit of Mother Earth

Smithsonian Guides to Natural America: The Northern Rockies
(with Thomas Schmidt)

Snow Country: Yellowstone in Winter
(with Steven Fuller)

Jeremy Schmidt is a writer and photographer specializing in natural science, conservation, and adventure travel. He is the author of more than a dozen books, and magazine feature stories for Audubon, International Wildlife, National Geographic, National Geographic Traveler, Natura (Italy), Panorama (Netherlands), Outside, GEO, and others.

His book *Himalayan Passage* won the first Barbara Savage Award for adventure writing. Other awards include the Lowell Thomas Award for travel writing (in National Geographic Traveler), and the Ranger Rick John Strohm Award. He founded the popular FreeWheeling Travel Guide series, and served for 12 years as the adventure columnist for Universal Press Syndicate. He is a founding faculty member of the Jackson Hole Writers Conference, and a Fulbright Scholar in the Council for International Exchange of Scholars.

He and his wife have lived in Jackson Hole, Wyoming for more than 25 years. His website is www.jerschmidt.com.